White Plymouth Rock Chickens of Quality and Usefulness
A Poultry Catalog of Maple Farm of Midlothian

by Charles T. Ettinger

with an introduction by Jackson Chambers

Self Reliance Books

Get more historic titles on animal and stock breeding, gardening and old fashioned skills by visiting us at:

http://selfreliancebooks.blogspot.com/

Introduction

I am pleased to present yet another title in the "Chicken Breeds" series.

This volume is entitled "White Plymouth Rocks of Quality and Usefulness". It was originally published in 1912 by Maple Farm of Midlothian as "Specialty Breeders of White Plymoth Rocks of Quality and Usefulness" and was an annual catalog issued by early poultryman, Charles D. Ettinger of Midlothian, Illinois near the City of Chicago.

Included are details on Mr. Ettinger's White Plymouth Rock Chickens, as well as his methods.

The work is in the Public Domain and is re-printed here in accordance with Federal Laws.

Though this work is a century old it contains much information on poultry that is still pertinent today.

As with all reprinted books of this age that are intended to perfectly reproduce the original edition, considerable pains and effort had to be undertaken to correct fading and sometimes outright damage to existing proofs of this title. At times, this task is quite monumental, requiring an almost total "rebuilding" of some pages from digital proofs of multiple copies. Despite this, imperfections still sometimes exist in the final proof and may detract from the visual appearance of the text.

I hope you enjoy reading this book as much as I enjoyed making it available to readers again.

Jackson Chambers

COUNTRY HOME OF CHARLES D. ETTINGER

EXECUTIVE OFFICE AND FARM HOUSE

MAIN HENNERY

INTRODUCTION

MAPLE FARM of Midlothian is situated eighteen miles south of Chicago, Illinois, on the Chicago, Rock Island & Pacific Railroad. Midlothian is our railroad, express and shipping station, and the farm lies two and one half miles west of the station. From May until November, during which time the Midlothian Country Club is open, the farm is reached by taking the Club car at Midlothian, which carries you up to the Club. From there the farm is almost due west and a ten-minute walk along the pavement brings you to Maple Farm of Midlothian.

From November until May, we will gladly meet anyone that wishes to come to the farm, with a conveyance at Midlothian, but we must insist that we be advised in advance, either by phone (our telephone number is Blue Island 362) or letter, of intended visits. This you must do to avoid an annoying wait at the station while the conveyance is getting there.

Tinley Park, Illinois, is our Post Office address, and Blue Island, Illinois, our telegraph office. This catalog is published in order to present to you as clearly as it is possible for us to do, our methods of doing business and to give in detail our plan of rearing top notch birds.

When you buy a bird, eggs for hatching or Baby Chicks you want all that you pay for. We intend to give every purchaser of poultry from this farm a square deal and we guarantee to live up to the rules of our catalog at all times.

Don't ask us to break our rules; you should know before you entrust us with your confidence and money what you will get for it. We guarantee to satisfy you on any purchase made here, but you should not expect a $15.00 bird for $5.00, and don't lay the blame on us if the lamp goes out in the brooder or incubator, or the hen leaves the nest.

We desire to use this medium to express to our friends our appreciation of their liberal patronage, and respectfully solicit a continuance of same.

We trust that all who read this book will find it interesting, and we cordially invite you to come and look our flock over and see for yourself that they substantiate our claims.

FIRST PRIZE PEN ILLINOIS STATE SHOW, JANUARY, 1912

MAPLE FARM A SPECIALTY FARM

IN EVERY sense you will find here a highly specialized farm. In the Poultry Department we breed White Plymouth Rocks exclusively. The White Plymouth Rocks of today can truly be said to be the best of the Rock family. And Maple Farm White Rocks are as high in quality as it is possible to produce anywhere. Picture the snow white plumage, the rich bay eye, the radiant face, intelligent head, grand yellow legs, superb shape and proud carriage, all combined in a bird of uniformity, and you have the Maple Farm White Rock.

Now when you consider that such a pleasing picture can produce eggs in quantities and have the vitality to produce chicks that will thrive and grow to be broilers with a rich yellow skin and plump breast at 8 and 9 weeks, and fancy soft roasters at from 16 to 20 weeks, and matured pullets ready to lay and reproduce the breed at from six to seven months, you have the fowl that will prove a profitable investment to you.

Every one of these qualifications is embodied in our White Rocks and we know that they will bear us out in our claims. If you try them you will be convinced.

THE POULTRY BUSINESS AS WE FIND IT

The poultry business is one where each and every operation is dependent upon the other. From the gathering of the large brown eggs to the finishing of the matured bird, it is necessary at all times to use judgment and care.

SHOWING A SECTION OF OUR SUNLIT BROODER HOUSE

Eggs should be gathered often. If left too long in the nest they may freeze, become cracked or broken by the hens, or the tender germ started by the heat of summer or the broody hen. All of these things are taken into consideration here at Maple Farm, and we gather them often enough to avoid all these contingencies. We keep all eggs in a room that very seldom rises above 60°, or drops below 45° in temperature.

We are trap-nesting our breeders from November until May, and in those seven months, the months when eggs are of the most value, we determine which birds to use in our breeding pens the following year. All birds that go into our trap-nest houses must first have the standard qualities of the type and breed. Then the pullets that show the best results for these seven months are kept over to use as yearling breeders the next season and their sons and daughters are recorded as chicks and go on to help raise the high standard of efficiency of our birds as layers. By this method we weed out the non-productive birds, so you can see that when buying Maple Farm birds you are buying birds that are tested, and proved by performance.

The incubators here are kept scrupulously clean and are thoroughly washed and disinfected after each hatch. The hens that are used for setting are set in clean, dry boxes in a quiet place, and at all times are kept free from lice. We practice the plan of taking each hen off of her nest every day up until the nineteenth day. In this way we are assured that they feed, drink and exercise sufficiently to keep them contented. And if the nest has been soiled or an egg broken we change the nesting and wash the eggs in warm water before putting the hen back on. These details all take time, but we realize the truth of the statement that the poultry business means doing a thousand and one little things well every day. And we have found by experience that we must do them to insure our success, which means your success when you handle Maple Farm White Rocks.

A SUMMER SCENE IN THE RUNS SOUTH OF BROODER HOUSE

THE BABY CHICK

FROM the time the chick is hatched we endeavor to make its life happy and its quarters comfortable. After the chicks come from the incubator they are moved into the brooder house, putting never over sixty to a brooder and generally only fifty. We use a hot water, fresh air system of brooding and the ease with which we raise them and the contentedness of their song tell us that our system is practical and dependable. The chicks are fed very sparingly the first few days. After they are seven or eight days old, they are given access to the runs, which are a few inches deep with good clover or alfalfa chaff or cut straw as litter. Fresh water is supplied them three times a day while they stay in the brooder house, and sufficient feed to keep them satisfied, but still hungry and active, always scratching and exercising their bodies, which condition promotes health and produces vitality.

THE MOTHER HEN

We let Nature have its way with the hen and her brood here, and only watch and see that they are contented and protected at all times.

After the chicks are dry in the nest, usually about the middle of the twenty-second day, we remove them to a clean individual coop, and after providing fresh water, we give the mother hen absolute quiet for twenty-four hours and allow her to tell her babies in her own way about the joys and disappointments of their life to come.

CHICKS ENJOYING THE UNLIMITED CLOVER RANGE

A SCENE ON ONE OF OUR CLOVER RANGES. SMALL PICTURE SHOWING DETAIL OF ONE OF THE HOUSES

The floors of these coops are covered one inch deep with clean, coarse sand.

After that we feed them much the same as we do the brooder chicks and allow them to stay with the hen until they are thoroughly feathered and able to look out for themselves. They are allowed plenty of range and at all times we are watching and waiting to protect them from the chill winds and storms and evil hawks.

A NEW WORLD FOR THE CHICKS

After the brooder chicks are five or six weeks old we move them out into Open Air Colony Houses that are provided with Colony Brooders. For the sake of those of you who have not seen these brooders, we will say just a word about their style and advantages. The houses are 6 x 10 feet in size and built on runners so they can be drawn about the farm with a team. The brooders consist of a small, slow-burning coal stove with an adjustable gal-vanized hover hung on pulleys from the ceiling with a balance weight. We have found that these brooders are entirely practical and after the chicks have been placed under them for a night or two they learn where the heat is, and readily go to it when cool.

Before allowing the youngsters to come out of the house we place a portable wire frame around the south side of the house. The chicks soon learn where the opening is, and after a week's time we can remove the frames entirely, allowing them unlimited range. Range affords liberty of action, and to grow chickens within the confines of a yard is to grow them with their active nature under restraint. On range the grass, insect life, worms and gravel satisfy the chicks' insatiable nature, and chicken growth and vitality are based on the law of satis-faction and contentment. After they are thoroughly feathered out, which is at about 10 weeks of age, we remove the brooders from the house. Here they

"ATTENTION, COMPANY!"

remain to go on and grow out to their maturity. Dry mash is kept before them at all times, as are grit, oyster shell and charcoal. Fresh water is so very essential to the chicks' normal, healthy growth, that we are always watchful that there is plenty of it and that it is clean. Time is precious, especially when the hatching season is on in full swing, so we employ the plan of watering all of the birds on the farm at noon during the open months. By having large fountains, ones that will hold more water than a given flock will drink in a day, we find that the water stays cooler and cleaner for a longer period by watering at noon; particularly cooler, because after the fountain is filled the volume of water helps to keep it cool, and then it cools off at night and is fresh for the chicks as soon as they get out in the morning. These are simple little plans that keep the chicks satisfied, and growing right, and at the same time save much labor.

WEANING THE CHICKS FROM THE HEN

Mother hen is a very wise bird, she will teach her chicks to eat and drink and scratch for worms, run to her when she calls either to give them a worm or to protect them from the showers; but she does not tell them what to do when they have left her for all time. So be careful; after we have weaned the chicks from the hen we put them in houses, 3 x 6 feet in size, on clover range. We place a portable wire frame in front of the house, so they may learn where the house is. After they have found this out, so that they get in out of a shower or at night by themselves, we remove the frame and allow them to range to their hearts' content. The care from then on is similar to the care which the incubator chicks receive. We have found, however, that it takes longer to train chicks weaned from hens to find their houses than it does incubator chicks. Bear that in mind; it will save lots of the little fellows.

FEEDING THE CHICKS

We use good, clean chick feed mixed with an equal part of rolled oats for all chicks until they are ten days old. Commencing at this age we give a morning feed of patented prepared chick food. This feed is semi-cooked and then dried and is moistened in water and then squeezed out before feeding. At four weeks we abandon the patented prepared food and start using a moistened, crumbly mash in the morning, containing bran, middlings, corn meal and rolled oats. The noon and night feeds are of mixed grain.

These feeds we continue using right along, increasing the amount as the chicks demand it.

At four weeks we commence feeding sprouted oats at noon. Grit, oyster shell and charcoal are kept before them constantly.

SUMMER DAYS

DURING the warm nights and hot days of summer every thought must be for the comfort and health of the young birds. An abundance of shade is absolutely necessary for their best growth and development, and too much stress cannot be laid on the value of pure, fresh water, clean houses, uncrowded floor space, and later uncrowded roosting space. Here at Maple Farm we have built almost every house off the ground, allowing enough room for the birds to get under during the hot days. Shrubs, trees and wooden shelters serve our purposes well in the open clover pastures, and allow the birds fine protection from the hot sun, at the same time allow the air to pass through at all times. We have found that the whitest birds will become tinged with brassiness if they haven't protection from the hot summer sun. Many a fine chick has been spoiled for showing in its first year by being subjected to the direct rays of the sun in summer.

THE WATER QUESTION

Water, pure, fresh and clean, is a most important item and one that we have found has a great deal to do with the general health of all stock. Our water comes from a driven well four hundred feet deep. It is pure, and has a very satisfactory effect on the birds—keeping them in excellent condition. The vessels which we use are kept free of scum and dirt, and every other day before refilling we use a brush in cleaning them to be sure to keep them free of any particles of dirt that may lodge on the side.

COCKERELS AND PULLETS ENJOYING A RED CLOVER PASTURE

A great many very promising chicks are ruined for life by being crowded in the houses at night. If they are moved away from the heat too soon, or their quarters are cold and drafty, they will pipe and squeeze against each other in a corner to keep warm, and generally where this condition exists you will find a number of dead chicks in the morning. We realize the utmost importance of plenty of fresh air and floor space here at Maple Farm, and our houses are so constructed that they are free from drafts but still allow more than enough fresh air to enter to keep an absolutely fresh supply for the chicks to breath, at all times. Up until the birds are old enough to roost we use clean torpedo sand on the floors of all colony houses, but after they commence to roost we use straw. We have found that by using sand it is very easy to go along with a fine rake and rake the droppings off every other day. This insures the health of the youngsters.

WHEN ROOSTING TIME COMES

After the youngsters are old enough to roost, we provide roosting poles made of two-by-fours, with the edges rounded off. These are laid flatwise, that is, the four-inch side up. This gives the youngsters whose bones are still very tender and impressionable, a wide space to rest their breasts on, and consequently it is the rarest thing for us to find a bird with a crooked breast bone. We always allow six inches of roosting pole space for every bird while they are young, and as they grow we increase it until we are allowing each bird twelve inches of space on the roosting perch. This gives the young birds plenty of room and prevents crowding and squeezing, which if allowed to happen will result in many narrow birds that are pinched in back.

PROPER FEEDING

Feed for the birds as they develop must, of course, be increased. Here at Maple Farm we always endeavor to keep the birds hungry, but at the same time we give them enough to keep them growing steadily.

Our birds on range receive a dry mash which is always before them. Moistened mash is fed in the morning, sprouted oats at noon and mixed grain at night.

These regular things watched and studied, season in and season out, have done as much as any other one thing to help us grow better birds—birds full of vitality, quality and usefulness.

WHITE PLYMOUTH ROCK MATING LIST

SEASON OF 1913

A MATED PEN. PRODUCERS OF MANY HIGH CLASS BIRDS

MAPLE FARM OF MIDLOTHIAN

BREEDERS OF

GUERNSEY CATTLE, CHESTER WHITE
SWINE, BELGIAN DRAFT HORSES AND
WHITE PLYMOUTH ROCK POULTRY

CHARLES D. ETTINGER, Proprietor   WALTER A. COOK, Manager

Mating List of 1913

At this season of the year is an opportune time to draw the attention of the poultry world to the question of eggs and baby chicks from which will come the layers and show birds of next fall and winter. MAPLE FARM OF MIDLOTHIAN, therefore, takes pleasure in announcing that during the hatching season of 1913 they will be in a better position than ever to supply their customers with the very cream of the White Plymouth Rock breed.

Although in a position to make up a large number of pens from which to sell exhibition eggs we have decided to make an extra effort to send out a higher quality of eggs than has ever before been attempted. Instead of mating up a dozen pens for high class exhibition stock, as has been our practice in the past, we have cut this down to four pens, two of hens and two of pullets. The very cream of our exhibition stock will go into these four pens and the remainder, including many birds ordinarily good enough for such matings, will go to make up our utility flocks.

Pen No. 1

This pen is headed by the cockerel that took 2nd prize at the Illinois State Fair held at Springfield, Ill., in October, 1912. He is a large, big-boned, typical White Plymouth Rock with a wonderfully well developed breast, broad back and deep body. His comb is ideal in shape and he has a fine wide head with a brilliant bay eye. Mated with this bird are ten hens that have been prominent winners at shows during the past two years. In this pen are two birds from the 1st pen at St. Louis, 1911, 2nd hen at Springfield, January, 1912, and two hens that were in the 2nd pen at Chicago, December, 1911, and the 1st pen at Springfield, January, 1912. The other birds are beautiful specimens that have never been shown.

Pen No. 2

At the head of this pen stands a cockerel that was 1st prize at the Illinois State Fair held at Springfield, Ill., October, 1912. At the time these birds were judged they were almost exact duplicates of each other, there being but very little difference in quality between these two birds. Mated with this cockerel are hens that have won the following prizes: one from the 1st pen at St. Louis, 1911; one from the 1st pen at Springfield, January, 1911; one from the 3rd pen at Chicago, December, 1911, as well as a bird that was 1st pullet at Springfield, January, 1912, and 2nd hen, Chicago, December, 1912. The other birds have been picked out with the idea of making them a well-balanced pen and we are looking forward to some wonderfully good chicks from this mating.

Pen No. 3

Pen No. 3 is made up of ten pullets mated with a cock bird. This cock bird is a sterling individual that headed 2nd pen at Chicago, December, 1911. He stood at the head of our pen No. 2 for our 1912 matings and we secured some very fine individuals from this bird. Mated with him are pullets that have been awarded the following prizes; two 5th pen pullets at Chicago, 1912; a 3rd prize bird at Springfield, October, 1912, and 5th prize bird at Chicago December, 1912, together with other females which are very excellent. This pen will undoubtedly raise some sterling individuals.

Pen No. 4

Pen No. 4 has been mated up carefully and contains birds that were in the 3rd pen at Springfield, October, 1912, and 5th pen at Chicago, December, 1912. Other pullets have been picked out from our large flock and mated with these prize birds until we have a pen here that is absolutely matched. This pen is headed by the bird that was 3rd prize cock bird at Chicago, December, 1911. He is certainly a fine individual of the White Plymouth Rock breed with plenty of breadth and length; his plumage is absolutely white. He has fine bay eyes, beautiful yellow shanks and an excellent comb. This mating should be ideal in every way.

Utility Pens

We have decided to use only a few birds for our exhibition matings, and we have, therefore, put the remainder of the show birds into our utility pens. Down along Colony Row, which is where our breeding houses are located, may be found birds that came from the 5th pen at Chicago, 1912, 3rd pen at Springfield, January, 1912, 3rd pen at Springfield, October, 1912, and one bird that was first prize at the large Illinois State Fair in the fall of 1912. Our utility matings are selected with the same care that is given our exhibition matings but necessarily do not contain as high type birds.

Eggs for Hatching

Eggs may be ordered from one pen or from several pens as the purchaser sees fit. Any setting that proves unsatisfactory may be returned to us and the infertile eggs will be replaced. All orders are filled in rotation so that if eggs are desired on any particular date the order should be booked early. We recommend that as soon as a customer receives a setting of eggs that he put them in a cool place where the temperature is as near 50 degrees Fahrenheit as possible and allow them to settle for at least 24 hours before putting them under the hen or the incubator.

Prices of Eggs

Our eggs from exhibition matings, that is, from pens 1 to 4 inclusive, are $10.00 per setting and $60.00 per hundred. The eggs from the utility pens are $3.00 per setting and $15.00 per hundred.

Baby Chicks

We have received a large number of inquiries for baby chicks during the past season and have promised a number of our customers that we will offer young chicks during the season of 1913. All orders shipped out will have the same close attention that has characterized our shipments in the past. Only large vigorous chicks will be sent and they will be packed in such a manner as to insure safe arrival. The prices on these will be $1.00 each or $75.00 per hundred from our exhibition matings, and 25c each or $20.00 per hundred from our utility matings.

"Maple Farm of Midlothian stands for the highest type of poultry that can be produced."

Our guarantee holds good in every department and we are willing that if a purchaser is not satisfied the stock may be shipped back, express prepaid, and we will gladly refund the money.

Such a guarantee is proof conclusive that we have the stock to deliver and are willing to let our customers be the judges. For any further particulars regarding our White Plymouth Rocks, write, or better still, pay a visit to

MAPLE FARM OF MIDLOTHIAN

Charles D. Ettinger, Proprietor

Walter A. Cook, Manager

Post Office: Tinley Park, Illinois.   R. R. Station: Midlothian, Illinois.

Telephone: Blue Island 362.

THE FINISHING STAGES

A S THE birds develop and are nearing maturity, it is very necessary to separate the cockerels from the pullets. The cockerels will consume more food than the pullets will at this stage in their lives, and unless separated the pullets will have trouble in getting all the food they need. At this time of separation we commence to go over the flock looking for weak specimens. Birds that don't show that alert, vigorous type and active sparkling eyes are discarded from the flock and fattened for table use. We have never shipped a bird from Maple Farm that didn't have an abundance of vitality, and never shall, but there are always a few that are, in the strict sense of the word, culls and our efforts are tireless to weed these birds out of the flock, as they most generally are unprofitable and never fit to go on and reproduce first-class birds.

About September 1st we start this separation and culling. From then on the pullets are put on an increased feed, in order to bring them to maturity, which means into laying marketable eggs as soon as possible.

We select the best of all the pullets to go into the laying houses first. Each and every one must possess vitality in abundance, above everything else. A proud, active carriage, an alert, sparkling eye, denote vitality, if anything does.

After we are sure that a specimen has this vitality, which is absolutely essential, we examine them for types and shape. The birds must have both of these characteristics or else they can't gain a position in our laying houses. Then we examine them for disqualifications, and of course, any disqualified bird is immediately discarded. Then the points of eye and shank and plumage color are considered.

SOUTH VIEW MAIN HENNERY. FINISHING HOUSES IN FOREGROUND

So, you see we are straining every effort to produce birds that will go on and reproduce birds that grow out better and better each year and strictly in accordance with the Standard of Perfection.

EXAMINING THE COCKERELS

Our examination of cockerels is much the same as with the pullets, and we are very sure that after we have culled our flock over the specimens to be found remaining in our finishing houses are going to develop into birds that will be worthy in every way, shape and form to go on and help build up the usefulness of Maple Farm White Rocks.

Right here, though, we wish to state that every bird that is grown out to maturity here is not a first prize winner. If they were, we would have to ask you much more for these high quality birds than we do.

As the cockerels develop they are watched and examined continually and extra fine specimens are removed to the conditioning rooms.

This insures protection from almost all accidents and puts us in a position to furnish specimens for almost any show on very short notice.

A recent visitor remarked, after spending several hours examining our flock and equipment, that he had never seen so many birds of such high class, even quality, and almost without exception in such perfect condition.

We spend a very large amount of time keeping our birds in a highly healthy condition. It isn't a hit or miss system that insures this, but a regular daily and weekly routine that is kept up at all times, winter or summer, so you can readily understand why our birds have made such an enviable reputation in the show room, not only for ourselves but for our customers as well. And with such care and attention you may rest assured that these same birds have the vitality and strength to prove profitable breeders and layers in any climate.

COCKERELS IN THE OUTSIDE FINISHING HOUSES

IN THE LAYING QUARTERS

AFTER the pullets have passed the rigorous test and examination which we submit them to, when we select the ones that are to occupy the winter laying houses, and they have been placed in their respective pens and leg banded, we are very careful to give these birds every attention that we can to insure their health and happiness, and at the same time get a large and satisfactory egg yield. Our houses are large and roomy, well ventilated and very light. We always allow five square feet of floor space for every bird, and have found by experience that the egg yield is larger when the birds have this much space than when they were allotted only three or four square feet of floor space.

A WORD ABOUT THE LITTER

During the winter we keep the floors covered with about fifteen inches of straw. This litter is very satisfactory in promoting exercise and consequently health when the birds have to scratch for their grain feeds.

THE WATER SUPPLY

A constant supply of fresh, clean water is kept before the fowls at all times, and in the winter we supply them with warmed water often enough so that it never freezes. The number of times a day that it is necessary to do this depends entirely on the temperature of the weather. This is a most important item, and one that takes considerable time, but when you stop to consider that an egg is composed of more than two thirds water, you will readily understand that the birds must have it if you wish a large egg yield, and it's the winter egg that brings the gold.

Cold, icy water is so ungrateful to the birds, that not only will they drink very little, but what they do drink takes up a very large amount of bodily heat to warm it, which means that a larger portion of the food that the bird consumes must go to keep up this heat instead of producing eggs. And again, it's eggs that you are after.

FEEDS AND FEEDING

The question of what to feed to get the golden winter eggs is one of unending interest; so many people, in widely separated localities, get results from almost entirely different rations, it is very hard to lay down any rules or formulas that will be infallible or practical in every locality.

However, there are a few simple basic rules that will help you to determine what to feed. The fowls must have food in combination that will supply the

First Cockerel
Illinois State Show
January, 1912

First Hen
Illinois State Show
January, 1912

First Cock
Illinois State Show
January, 1912

First Pullet
Illinois State Show
January, 1912

constituents of which the egg is composed. Then it must contain muscle and vitality producing elements, as well as fat to keep the body warm and in good flesh, and the same time these three functions are being considered it is necessary to remember that the combination must be palatable, and it will be further necessary to change the proportions every six or eight weeks in order to keep the appetite sharp. Just what the ration should consist of will depend upon the availableness of the different ingredients and the cost of same.

Our basic mash ration consists of two parts bran and one part each of corn meal, wheat middlings, rolled oats, alfalfa meal and beef scrap, and one half part each of gluten meal and cotton seed meal, with 3 per cent fine charcoal and 2 per cent fine salt added to the total weight of the other feeds.

We shift these ingredients about, changing the quantities every six or eight weeks, and in the moulting season we add about 15 per cent linseed oil meal.

FEEDING THE MASH

We keep hoppers filled with dry mash before the birds constantly, and give them a feed in the morning of hot moistened mash besides. We use pumpkins, mangelwurzels, sugar beets and small potatoes along with the mash. These are boiled thoroughly and the whole mixed together with the mash to a crumbling consistency, making a fine hot feed for the birds which they eat with great relish.

READY FOR BUSINESS IN FRONT OF ONE OF THE WINTER LAYING PENS

At noon we feed either buttermilk, table scraps or sprouted oats, and at night a feed of mixed grain is scattered into the litter. This keeps the birds busy up until roosting time, and gives them something to scratch for and warm up their bodies on the first thing in the morning.

This ration is composed of two parts each of whole wheat and large, sifted white cracked corn, and one part each of kaffir corn, rolled oats, barley and buckwheat. This grain ration is also changed to give variety, and according to seasons of the year and the consequent economic values of the different grains.

THE CARE OF THE PENS

The droppings are removed from the boards every other day and the straw is changed often enough to keep it fresh, clean and sweet. Many diseases that the chicken is heir to are directly attributable to musty, foul-smelling litter. So be sure to keep your litter clean.

Every three months the houses are thoroughly cleaned and sprayed with whitewash that has about 5 per cent of crude carbolic acid added to it, and as a result of these practices we very seldom have any sick fowls, and never have we had an epidemic of any disease on our plant.

INTERIOR OF ONE OF THE LAYING PENS

THE BREEDING BIRDS

HERE at Maple Farm females that are in our breeding pens have earned their places by meritorious work and by their type and shape, being as near to the Standard of Perfection as we can breed them, and that we are breeding many high class birds of wonderful evenness is proved by our show records. Last year at St. Louis in a very large and fine class of birds we won **First Pen, First** and **Fourth Hen** and **Fourth Pullet,** also the **White Rock Club Ribbons for Best Pen and Hen.** At Chicago, December, 1911, we controlled **Second and Third Pen Birds** and also **Third Cock.** At the **Illinois State Show,** Springfield, Illinois, January, 1912, we won **every First and Special offered.** Our total winnings there were as follows:

First and Fourth Cock. First, Second and Third Hen. First and Third Cockerel. First and Second Pullet. First and Third Pen.

First Prize of $100.00 for the best and largest display, besides several cups and other specials.

This fall at the **Illinois State Fair,** October, 1912, where, on showing only young birds, we won, **First, Second, Fourth and Fifth Cockerel; First, Third and Fifth Pullets; and Third Pen.**

LOOKING DOWN "BREEDERS' ROW"

These winnings stand out most prominently and speak volumes for the quality of our birds. These grand birds, together with hundreds of other birds of equal merit and evenness, composed our matings for the season of 1912.

As a result, that we have already proven the worth of these matings is shown by our recent October winning, and right here we wish to say that we have several hundred birds maturing here daily that are high class specimens of sterling worth.

Our trap-nest records tell us which birds are our very best layers and by breeding only from those that are large producers and most excellent in every line, section and characteristic of body, we have attained an average that is very high.

We are constantly discarding birds from our breeding pens that don't toe the mark, or that develop some disqualification, thereby reducing the chances of raising any chicks that will develop into culls, or of sending out any baby chicks, eggs or grown birds that will not be satisfactory specimens to go on and help us build still higher our reputation for honest value and square dealing.

These birds are fed and handled much the same as our laying birds, the only difference being that they are mated into pens of from ten to twelve females and one male.

Constant attention to details, an ever-vigilant eye to the needs and comforts

A MATED PEN. PRODUCERS OF MANY HIGH CLASS BIRDS

INTERIOR OF ONE OF THE BREEDING HOUSES

of our birds, clean houses, regular feeding and watering have been the backbone of our success.

So you may feel satisfied that the birds as quoted herewith are honest values for the money and have been raised in a clean, sanitary manner, that produces birds of quality that have vitality, and blue-blooded egg records in their blood lines.

PRICES OF MATURED STOCK

EXHIBITION BIRDS

W E ARE in a position to furnish finished trained birds ready for the show room at all times. If you will advise us what show you intend to make, giving us all the information you can about the show, such as number of birds in previous shows, etc., and tell us what you are willing to spend, we will write you explicitly, telling you just what we can furnish for the money.

If you wish us to guarantee the Blue Ribbon for you, we will make you a definite price and if by any chance our birds fail to capture the coveted prize, we make you a sliding scale proposition. This is always a matter of correspondence.

SELECTED BREEDERS

When you study our prices for this class of birds, we wish you to bear in mind that our selected breeders are all first-class birds, absolutely free from any disqualifications, and all possessing the characteristics of the breed as called for in the Standard of Perfection, and in addition, remember that our birds carry a wealth of vitality and are proved and tested as producers.

You will find that they will prove a most profitable investment. They will fill your egg basket with lots of fine, large, brown eggs, and they will stamp

A SECTION OF OUR CONDITIONING ROOM

their quality in every way on their offspring. At the different prices quoted on this page you naturally will expect a graduating scale of excellence.

At these different prices we will give you more than full value for your money and guarantee to satisfy you.

	$ 7	$10	$12	$15	$ 25	$ 40	$ 50
Cocks	$ 7	$10	$12	$15	$ 25	$ 40	$ 50
Hens	4	6	8	10	12	15	
Cockerels	4	6	8	10	15	20	25
Pullets	3	5	7	10	12	15	
Trios	10	11	12	13	15	20	25
Breeding Pens							
Five females and one male		20	25	35	50	75	100
Ten females and one male	30	40	50	75	100	150	200

GENERAL DESCRIPTION OF THESE BIRDS
AS PRICED ABOVE

COCKS

$7.00. At this price we will furnish a good-sized bird, one with good shape and type and white. Many good chicks may be expected from him.

$10.00. A strong, vigorous, attractive bird. Good size, comb nice, full breast and very white. A bird sure to stamp his quality on his chicks.

BREEDERS DISPLAY, CLUB AND ASSOCIATION CUPS WON AT ILLINOIS
STATE SHOW, JANUARY, 1912

$12.00. Similar to the $10.00 bird, but better in the minor points, such as eye, color, comb, shank, etc.

$15.00. A very fine cock bird you will get at this price. A bird good in all sections and fully qualified to go into the show room.

$25.00. Will buy a cock bird of very fine quality. A bird sure to be valuable as a breeder and will hold his own in the show room. Deep, broad backed, full round chest, neat low comb, fine bay eye, excellent bone, grand shank and beak color and dead white.

$40.00. Buys one of our tried and proved breeding cocks. Such a bird will be exceptionally fine in all sections. Quantities of vitality, with all the standard qualifications and characteristics; a winner in the show room of most any show.

$50.00. Here we will send you an absolutely top notch bird. Every section fine and true to type and possessing that grand, alert carriage, full of vigor with a sparkling, red eye; the kind that have made Maple Farm White Rocks winners wherever they go. This bird will be a delight for his owner at all times.

COCKERELS

You will notice our cockerel prices are just about half of what we ask you for cock birds. This is because we don't have to carry them over for a whole year. You may expect at the different prices birds just as good in every particular as the cock birds. We figure that a cockerel worth $25.00 at 8 or 9 months of age is worth $50.00 at 18 to 20 months. Reason it out for yourself.

HENS AND PULLETS

Our pullet prices you will notice are slightly less than our prices for hens. This is true as with cockerels and cocks. The only reason there isn't as great a disparity in the prices is because our pullets pay their way all the time they are becoming hens, and when they reach their maturity, if they have passed our rigid tests, they are easily worth the difference.

$3.00 Pullets, $4.00 Hens.

At these prices we will furnish you a very choice female, one that is good in all sections and white. A bird like this carries our blood lines of vitality and usefulness, which will show in her offspring.

$5.00 Pullets and $6.00 Hens.

We will send you birds at these prices that will be very choice in every section. Such a hen would be one that we have used in our breeding pens and the pullet will be equal in every particular.

$7.00 Pullets, $8.00 Hens.

At these prices we will furnish extra choice birds—birds with an abundance of vigor, grand in type and shape, very white and very choice in minor points.

Such birds will breed many excellent high class specimens and if properly fitted will stand well up in the show room in reasonable competition.

$10.00 Pullets and Hens.

At this price we value our pullets as highly as hens, because such a bird will be equally choice in every section. Birds of this price carry our very highest quality in their breeding and will be most excellent specimens all through.

$12.00 Pullets and Hens.

This extra margin over the $10.00 birds will insure your getting a bird that is a little smoother in all the fine points. And the extra investment will be more than returned in the transmission of these points on the offspring. These finer points consist of better eye, comb and shank color and more perfect type and shape.

$15.00 Pullets and Hens.

These birds are top notch specimens in every way. The hens will come from our tried and proved breeding pens. One that is grand in all sections and sure to give a most excellent account of herself in the show room, as well as producing a wealth of high class chicks. The pullets at this price are equal to the hens. Those large, broad backed, long deep bodied birds, which lay so many eggs and produce so many excellent chicks.

TRIOS

A $10.00 Trio consists of two $4.00 hens and one $4.00 cockerel.

An $11.00 Trio consists of one $7.00 cock bird and two $3.00 pullets.

A $12.00 Trio consists of one $6.00 cockerel and two $4.00 hens.

A $13.00 Trio will consist of a $7.00 cock bird and two $5.00 pullets.

A $15.00 Trio will consist of an $8.00 cockerel and two $6.00 hens.

A $20.00 Trio will consist of one $10.00 cock bird and two $7.00 pullets. They will breed many winners too.

A $25.00 Trio will consist of a $15.00 cockerel and two $8.00 hens.

SMALL BREEDING PENS

For $20.00 we will send you five $3.00 pullets and a $7.00 cock bird.

For $25.00 we will supply five of our regular $4.00 hens and an $8.00 cockerel. A fine flock to start with.

At $35.00 we will send you five of our $5.00 pullets and a $15.00 cock bird. This pen will surely be a paying investment.

At $50.00 we will furnish five of our regular $8.00 hens and a $15.00 cockerel. Such quality as this pen will contain will surely produce excellent show birds.

At $75.00 we will ship you five $12.00 pullets and a $25.00 cock bird. Here's a bargain that will more than satisfy you in every way.

At $100.00 we will ship you either five of our $15.00 pullets or hens and a $50.00 male bird.

We are positive that any of these birds will score in almost any competition and produce quantities of exhibition specimens. Their splendid type and shape and superb carriage and even quality will be a delight to you at all times.

LARGE BREEDING PENS

$30.00 invested here will bring you ten $3.00 pullets and a $7.00 cock bird.

A $40.00 pen will consist of ten $4.00 hens and an $8.00 cockerel.

For $50.00 we will send you ten grand $5.00 pullets and a $10.00 cock bird. Here's a winner.

For $75.00 we will send you a pen that's an excellent value, ten $6.00 hens and one $25.00 cockerel.

For $100.00 invested here you will receive ten $7.00 pullets and a $40.00 cock bird.

For $150.00 we will send you ten grand show hens, birds that we receive $12.00 apiece for singly, and a large grand show cockerel, such as we ask $40.00 for and guarantee to win a blue for you.

For $200.00 we will ship you ten excellent pullets, our regular $15.00 quality, and a first-class cock bird, one almost good enough to win any show in the country. Such a bird as we ask $100.00 for and guarantee him to win.

These pens are all mated and will produce quantities of chicks overflowing with sterling points of quality.

BABY CHICKS

We have had so many demands for baby chicks that we have decided to offer them for sale the coming season.

Our chicks will surely prove very satisfactory and we take the utmost care in packing and delivering to the express office.

Our hatching is done in clean, sanitary incubators, and the vitality of the breeding stock insures our delivering chicks that will live and grow to be healthy specimens of the breed. We offer chicks from both our exhibition and utility matings and guarantee to give you what you pay for.

Exhibition chicks . . . . $1.00 each, $75.00 per 100
Utility chicks . . . . .25 each, $20.00 per 100

EGGS FOR HATCHING

We have often said that we would much rather never sell an egg for hatching. The uncertainty of the outcome is usually very embarrassing if the hatch is a poor one, and especially if the purchaser is a beginner, and does not fully appreciate the fact that although the eggs we send out are identical with the ones we set and have success with here, his eggs may not hatch very well, due to no fault of ours.

But to those who want eggs we will gladly supply them and guarantee their safe arrival and will replace any setting that proves unsatisfactory at hal price if the purchaser will return us the infertile eggs express prepaid.

PRICES OF EGGS

Our Exhibition Pen Mating Eggs are $10.00 per sitting, $30.00 per 50, and $60.00 per 100.

Our Utility Pen Eggs are $3.00 per sitting and $15.00 per 100. All sittings consist of fifteen eggs.

A CORNER OF THE INCUBATOR CELLAR

GRANARY AND MILL

LOOKING DOWN GUERNSEY AVENUE

A CORNER IN THE PACKING ROOM

MAIN OFFICE

OUR REFERENCES AND TERMS

Before entrusting us with your valued patronage, if there is any doubt in your mind as to the reliability of Maple Farm, we request that you write the Continental & Commercial National Bank of Chicago, Illinois, or any poultry journal in which our advertisement appears.

We want you to feel absolutely sure that you will get full value for your money. We guarantee to satisfy you.

FACILITIES FOR SHIPPING

Being centrally located, as we are, only eighteen miles from Chicago, the greatest railroad centre of the country, gives us the advantage of shipping promptly to any part of the world. Four express trains a day pass Midlothian en route to Chicago, and there connections can be made with almost every railroad in the United States, reaching to every corner, and fast express trains to every port for foreign shipments.

If you favor us with an order, please bear in mind that we have a personal interest in your success. Our honest endeavor is to make you a satisfied customer and we will do anything within the bounds of reason to accomplish this end. Wishing you a most successful and profitable season, we beg to remain, Respectfully yours,

MAPLE FARM OF MIDLOTHIAN
CHARLES D. ETTINGER

A SHIPMENT ON THE ROAD TO ITS DESTINATION